C000118103

BABY STEPS IN PHYSICS

Free-Fall and Projectile Motion

Problems in Physics with Solutions

by Boris Sapozhnikov

Web site: http://How2Physics.com

Facebook page: https://www.facebook.com/How2Physics

Twitter: @How2Physics or https://twitter.com/how2physics

ISBN: 978-0-9948028-4-2

I acknowledge, my wife, Faina Sapozhnikov, who has supported me in all of my endeavors and taken care of our children and me.

CONTENTS

Part One. Problems.

Free-Fall. A Vertically Launched Projectile.

1. An object is dropped (released from rest) from a high building.

   a) Find the velocity of the object after $5 \; seconds$.

   b) Find the displacement of the object during the first $5 \; seconds$.

2. An object is dropped from the top of a building and hits the ground $4 \; seconds$ from release.

   a) How tall is the building?

   b) Find the position of the object as a function of time.

   c) Find the velocity of the object when it hits the ground.

3. A ball is thrown straight up with an initial velocity of $15 \; \frac{m}{s}$.

   a) Find the position of the ball as a function of time if it is released from a height of $1.5 \; m$.

   b) How much time does it take for the ball to reach its maximum height?

   c) How long was the ball in the air?

   d) How high will the ball go?

4. A stone is thrown straight up from the top of a building with an initial velocity $10\ \frac{m}{s}$ and hits the ground $5\ seconds$ later.

    a) Find the height of the building .

    b) Find the position of the stone as a function of time.

    c) Simultaneously, another stone is thrown straight down with a velocity $10\ \frac{m}{s}$. Which of the two stones will hit the ground with a large velocity?

5. An object is thrown upwards from the ground with an initial velocity of $20\ \frac{m}{s}$. Two seconds later another object is thrown upwards with an initial velocity of $30\ \frac{m}{s}$.

    a) Find the position of the first object as a function of time.

    b) Find the position of the second object as a function of time.

    c) Determine when and where they will meet.

    d) Draw a position-time graph of the objects' motion.

6. One stone is thrown upwards from a point on the ground with an initial velocity $20\ \frac{m}{s}$ and simultaneously another stone is dropped from a height. They reach the ground at the same instant.

    a) Compare the velocities which they have attained.

    b) Find the height from which the second stone was dropped.

    c) Draw a velocity-time graph of the stones' motion.

7. A hot air balloon is travelling vertically upwards with constant velocity of $5 \frac{m}{s}$. A sandbag is released and it hits the ground $12 \ seconds$ later.

    a) With what speed does the sandbag hit the ground?

    b) How high was the balloon when the sandbag was released?

    c) How high will the sandbag go?

    d) If we assume that the balloon's velocity increased to $5.5 \frac{m}{s}$ after releasing the sandbag, what is the relative velocity of the sandbag with respect to the balloon $6 \ seconds$ after it was dropped?

8. A stone is thrown straight up from the surface of a planet without atmosphere. The graph shows the velocity as a function of time for the stone.

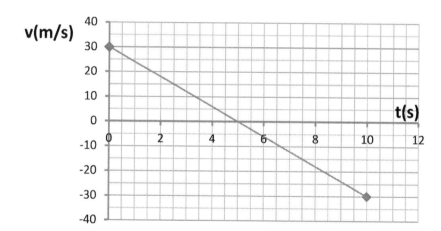

    a) What is stone's initial velocity?

    b) What is the magnitude of the free-fall acceleration on the planet?

    c) How high will the stone go?

    d) Draw an acceleration-time graph of the stones' motion.

3

A Horizontally Launched Projectile.

9. An object is dropped from an airplane flying horizontally at $125 \frac{m}{s}$. The object falls for 18.6 s before it hits the ground.

    a) From what height was the object dropped?

    b) What is its vertical velocity when it strikes the ground?

    c) How fast is it travelling horizontally when it strikes the ground?

    d) What is the object's velocity when it strikes the ground?

    e) How far did the object fall downwards during the fifth second after being released?

10. A tennis ball is served horizontally at a speed of $24 \frac{m}{s}$ from a height of 2.7 m. The net is 1 m high and 12 m horizontally from the server.

    a) How long does it take for the ball to cover the distance between the server and the net?

    b) Determine whether the ball clears the net and if so by what distance.

11. Two vertical towers stand on horizontal ground and are of heights 50 m and 40 m. A stone is thrown horizontally from the top of the highest tower with a velocity of $25 \frac{m}{s}$ and just clears the smaller tower.

    a) Find the distance between the two towers.

    b) Find the distance between the smaller tower and the point on the ground where the stone lands.

    c) Draw a trajectory (flight path) of the stone.

12. Two vertical towers stand on horizontal ground and are of heights $4h$ and h. The horizontal distance between the towers is X. When a stone is thrown horizontally with velocity v from the top of the taller tower towards the smaller tower, it lands at a point $\frac{3}{4}X$ from the taller tower. When a stone is thrown horizontally with velocity u from the top of the smaller tower towards the taller tower, it also lands at the same point.

Prove that $3u = 2v$.

13. A bomber flies with a constant velocity of $50\ \frac{m}{s}$ horizontally and wants to hit a target traveling $20\ \frac{m}{s}$ (same direction) on a highway $1000\ m$ below.

    a) What is the horizontal distance of the bomber from the target so that a bomb released from it will hit the target?

    b) At what angle (with the horizontal) should the target be in the pilot's sights when the bomb is released?

    c) Calculate the velocity (magnitude and direction) with which the bomb hits the target?

14. An apple is thrown horizontally from the top of a building $75\ m$ high with an initial velocity $10\ \frac{m}{s}$. One second later an arrow is shot vertically upwards from ground level with an initial velocity of $30\ \frac{m}{s}$. The arrow hit the apple.

    a) What is the horizontal distance between the building and an archer?

    b) Calculate the velocity (magnitude and direction) of the apple and the arrow immediately before the impact.

A Projectile Launched At An Angle.

15. A soccer ball is kicked from the ground with an initial velocity of $20\ \frac{m}{s}$ directed 40^0 above the horizontal.

   a) Find the horizontal position of the soccer ball as a function of time.

   b) Find the vertical position of the soccer ball as a function of time.

   c) Find the equation of the trajectory.

   d) How long was the soccer ball in the air?

   e) At what distance from the initial point will the ball hit the ground?

16. Ten seconds after being fired, a projectile strikes a hillside at a point displaced $700\ m$ horizontally and $90\ m$ vertically above the point of launch.

   a) Calculate for the projectile its initial velocity (magnitude and direction).

   b) Calculate for the projectile its maximum height above its launch point.

   c) Calculate for the projectile its velocity (magnitude and direction) just before impact.

17. Prove that

   a) the horizontal range of a projectile is given by $R = \frac{v_i{}^2 \cdot \sin(2\alpha)}{g}$.

   b) for a given value of initial velocity, the horizontal distance (range) would be greatest at a launch angle of 45^0.

   c) the maximum height of a projectile is given by $H = \frac{v_i{}^2 \cdot \sin^2(\alpha)}{2 \cdot g}$.

18. A person can throw a stone to a maximum horizontal distance of $100\ m$.

   a) Calculate the initial velocity of the stone.

   b) What is the maximum height that the stone reaches?

   c) How high can he throw the same stone?

19. An airplane releases a package at an instant when the airplane has a velocity of $80\ \frac{m}{s}$ at an angle of 30^0 below the horizontal. At this instant of release, the airplane is 1200 m above level ground.

   a) How much time elapses between the release and the impact with the ground?

   b) How far does the package travel horizontally after release?

   c) What is the velocity of the package just before impact?

   d) Draw horizontal position-time and vertical position-time graphs showing the motion of the package.

20. As a ship is approaching the dock at $1\ \frac{m}{s}$, a package needs to be thrown to it before it can dock. The package is thrown at $17\ \frac{m}{s}$ at 50^0 above the horizontal from the top of a tower at the edge of the water, $10\ m$ above the ship's deck. The package should land at the front of the ship.

   a) How long was the package in the air?

   b) What is the velocity of the package just before impact?

   c) At what distance from the dock should the ship be when the package is thrown?

   d) Draw horizontal velocity-time and vertical velocity-time graphs of the package.

Part Two. Solutions.

Free-Fall. A Vertically Launched Projectile.

We assume that the only force acting on an object is the force of gravity which causes it to accelerate towards the center of the Earth at a constant rate of $9.8 \frac{m}{s^2}$. Also, we refer the motion to the vertical y axis with +y vertically up. Consequently, the acceleration has to be negative, the velocity is positive when the object is rising and is negative when the object is falling.

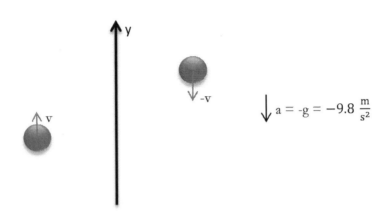

1.

a)

The equation for velocity as a function of time is given by

$$v_f = v_i + a \cdot t$$

where v_f - the final velocity of the object, v_i – the initial velocity of the object, a – the acceleration and t – the time it takes the object to change its velocity.

We rewrite the equation for velocity as a function of time by setting $v_i = 0$ (the object is dropped) and $a = -g$ (the acceleration due to gravity).

$$v_f = -g \cdot t$$

We know that $-g = -9.8 \frac{m}{s^2}$ and $t = 5\ s$.

Hence,

$$v_f = -g \cdot t = -9.8 \cdot 5 = -49 \frac{m}{s}$$

Note: The minus sign shows the direction of the velocity.

b)

The position-time equation of motion of the object is given by

$$y_f = y_i + v_i \cdot t + \frac{a \cdot t^2}{2}$$

where y_f - the final position of the object, y_i – the initial position of the object, v_i – the initial velocity of the object, a – the acceleration and t – the time it takes the object to move from the initial position to the final position.

We rewrite the position-time equation of motion of the object by setting $v_i = 0$ and $a = -g$.

$$y_f = y_i - \frac{g \cdot t^2}{2}$$

Displacement is the shortest distance between initial and final positions and is given by the formula

$$\Delta y = y_f - y_i$$

where Δy – the displacement.

Hence,

$$\Delta y = -\frac{g \cdot t^2}{2}$$

We substitute the values into the equation and solve for Δy

$$\Delta y = -\frac{g \cdot t^2}{2} = -\frac{9.8 \cdot 5^2}{2} = -122.5 \text{ m}$$

Note: The minus sign indicates that the displacement is in the downward direction.

2.

a)

The position-time equation of motion of the object is given by

$$y_f = y_i + v_i \cdot t + \frac{a \cdot t^2}{2}$$

where y_f - the final position of the object, y_i – the initial position of the object, v_i – the initial velocity of the object, a – the acceleration and t – the time it takes the object to move from the initial position to the final position.

We rewrite the position-time equation of motion of the object by setting $v_i = 0$ and $a = -g$.

$$y_f = y_i - \frac{g \cdot t^2}{2}$$

Displacement is the shortest distance between initial and final positions and is given by the formula

$$\Delta y = y_f - y_i$$

where Δy – the displacement.

Hence,

$$\Delta y = -\frac{g \cdot t^2}{2}$$

We substitute the values into the above equation and solve for Δy

$$\Delta y = -\frac{g \cdot t^2}{2} = -\frac{9.8 \cdot 4^2}{2} = -78.4 \text{ m}$$

In our case, the absolute value of the displacement is the height of the building 78.4 m.

b)

We use the position-time equation of motion of the object to write the position of the object as a function of time

$$y_f = y_i + v_i \cdot t + \frac{a \cdot t^2}{2}$$

where y_f - the final position of the object, y_i – the initial position of the object, v_i – the initial velocity of the object, a – the acceleration and t – the time it takes the object to move from the initial position to the final position.

We rewrite the position-time equation of motion by renaming y_f by y and by setting $v_i = 0$ and $a = -g$.

$$y = y_i - \frac{g \cdot t^2}{2}$$

We know that $-g = -9.8 \, \frac{m}{s^2}$ and the initial position is 78.4 m (an object is dropped from the top of a building).

Hence,

$$y = y_i - \frac{g \cdot t^2}{2} = 78.4 - \frac{9.8 \cdot t^2}{2}$$

or

$$y = 78.4 - 4.9 \cdot t^2$$

c)

The equation for velocity as a function of time is given by

$$v_f = v_i + a \cdot t$$

where v_f - the final velocity of the object, v_i – the initial velocity of the object, a – the acceleration and t – the time it takes the object to change its velocity.

11

We rewrite the equation for velocity as a function of time by setting $v_i = 0$ (the object is dropped) and $a = -g$ (the acceleration due to gravity).

$$v_f = -g \cdot t$$

We know that $-g = -9.8 \, \frac{m}{s^2}$ and $t = 4 \, s$.

Hence,

$$v_f = -g \cdot t = -9.8 \cdot 4 = -39.2 \, \frac{m}{s}$$

3.

a)

The position-time equation of motion of the object is given by

$$y_f = y_i + v_i \cdot t + \frac{a \cdot t^2}{2}$$

where y_f - the final position of the object, y_i – the initial position of the object, v_i – the initial velocity of the object, a – the acceleration and t – the time it takes the object to move from the initial position to the final position.

We use the position-time equation to write the position of the ball as a function of time. Renaming y_f by y and setting $a = -g$, we get

$$y = y_i + v_i \cdot t - \frac{g \cdot t^2}{2}$$

We know that the initial position is $1.5 \, m$, the initial velocity of the object is $15 \, \frac{m}{s}$ and the acceleration is $-g = -9.8 \, \frac{m}{s^2}$.

Then,

$$y = 1.5 + 15 \cdot t - \frac{9.8 \cdot t^2}{2}$$

and finally,

$$y = 1.5 + 15 \cdot t - 4.9 \cdot t^2$$

b)

The equation for velocity as a function of time is given by

$$v_f = v_i + a \cdot t$$

where v_f - the final velocity of the object, v_i — the initial velocity of the object, a — the acceleration and t — the time it takes the object to change its velocity.

In addition, we can use the fact that the object will come to a complete stop for an instant when it reaches its maximum height.

So, we rewrite the equation for velocity as a function of time by setting $v_f = 0$ (the object will come to a complete stop) and $a = -g = -9.8 \frac{m}{s^2}$.

$$0 = v_i - g \cdot t$$

or

$$0 = v_i - 9.8 \cdot t$$

and finally,

$$t = \frac{v_i}{9.8}$$

Now we substitute the value of the initial velocity into the formula and solve for t.

$$t = \frac{15}{9.8} \approx 1.53 \text{ s}$$

c)

We know the position of the ball as a function of time

$$y = 1.5 + 15 \cdot t - 4.9 \cdot t^2$$

Using this equation and the fact that the ball will reach the ground $y = 0$, we will get

$$1.5 + 15 \cdot t - 4.9 \cdot t^2 = 0$$

Solving this quadratic equation for t, and keeping only the physically meaningful positive root (the time interval is always positive), we then have

$$t \approx 3.16 \text{ s}$$

d)

The position of the ball as a function of time is given by the formula

$$y = 1.5 + 15 \cdot t - 4.9 \cdot t^2$$

Also, we know that it takes 1.53 s for the ball to reach its maximum height.

Then,

$$y = 1.5 + 15 \cdot 1.53 - 4.9 \cdot 1.53^2$$

and finally,

$$y = 12.98 \ m$$

4.

a)

The position-time equation of motion of the stone is given by

$$y_f = y_i + v_i \cdot t - \frac{g \cdot t^2}{2}$$

where y_f - the final position of the stone, y_i – the initial position of the stone, v_i – the initial velocity of the stone, g – the acceleration due to gravity and t – the time it takes the stone to move from the initial position to the final position.

We know that the final position is $0 \ m$ (the stone hits the ground), the initial position is h (the height of the building), the initial velocity of the stone is $10 \ \frac{m}{s}$, the time it takes the stone to reach the ground is $5 \ s$ and the acceleration is $9.8 \ \frac{m}{s^2}$.

14

Then,

$$0 = h + 10 \cdot 5 - \frac{9.8 \cdot 5^2}{2}$$

and finally,

$$h = 72.5 \ m$$

b)

We use the position-time equation to write the position of the stone as a function of time.

$$y_f = y_i + v_i \cdot t - \frac{g \cdot t^2}{2}$$

We know that the initial position is $y_i = h = 72.5 \ m$, the initial velocity of the stone is $10 \ \frac{m}{s}$ and the acceleration is $g = 9.8 \ \frac{m}{s^2}$.

Thereby,

$$y = 72.5 + 10 \cdot t - \frac{9.8 \cdot t^2}{2}$$

or

$$y = 72.5 + 10 \cdot t - 4.9 \cdot t^2$$

Notice that we denote y_f by y for convenience.

c)

First, we find the velocity of the first stone when it hits the ground.

The equation for velocity as a function of time is given by

$$v_f = v_i - g \cdot t$$

where v_f - the final velocity, v_i – the initial velocity, g – the acceleration due to gravity and t – the time it takes to change its velocity.

We substitute the values into the equation and solve for v_f

$$v_f = 10 - 9.8 \cdot 5 = -39 \, \frac{m}{s}$$

Now we can find the velocity of the second stone when it hits the ground. We use the velocity-position equation of motion.

The velocity-position equation of motion of the object is given by

$$v_f{}^2 = v_i{}^2 + 2 \cdot a \cdot (y_f - y_i)$$

where y_f - the final position of the object, y_i – the initial position of the object, v_f - the final velocity of the object, v_i – the initial velocity of the object and a – the acceleration.

We rewrite the velocity-position equation of motion of the object by setting $a = -g$.

$$v_f{}^2 = v_i{}^2 - 2 \cdot g \cdot (y_f - y_i)$$

We substitute the values into the equation and solve for v_f

$$v_f{}^2 = (-10)^2 - 2 \cdot 9.8 \cdot (0 - 72.5)$$

$$v_f = \pm 39 \, \frac{m}{s}$$

We know that the direction is negative, so that we have to choose the negative sign.

$$v_f = -39 \, \frac{m}{s}$$

Thereby, both stones will reach the ground with the same velocity.

5.

a)

The position-time equation of motion of the object is given by

$$y_f = y_i + v_i \cdot t - \frac{g \cdot t^2}{2}$$

16

where y_f - the final position of the object, y_i – the initial position of the object, v_i – the initial velocity of the object, g – the acceleration due to gravity and t – the time it takes the object to move from the initial position to the final position.

We know that the initial position is $0\ m$ (The object is thrown upwards from the ground.), the initial velocity of the object is $20\ \frac{m}{s}$ and the acceleration is $9.8\ \frac{m}{s^2}$.

Then,

$$y_1 = 0 + 20 \cdot t - \frac{9.8 \cdot t^2}{2}$$

or

$$y_1 = 20 \cdot t - 4.9 \cdot t^2$$

Notice that we denote y_f by y_1.

b)

The position-time equation of motion of the second object is given by

$$y_f = y_i + v_i \cdot t - \frac{g \cdot t^2}{2}$$

We know that the initial position is $0\ m$ (We assume that the second object is thrown upwards from the ground.), the initial velocity of the second object is $30\ \frac{m}{s}$ and the acceleration is $9.8\ \frac{m}{s^2}$.

Also, we can write that the time is $(t - 2)$ because the second object was thrown two seconds later.

Then,

$$y_2 = 0 + 30 \cdot (t - 2) - \frac{9.8 \cdot (t - 2)^2}{2}$$

or

$$y_2 = 30 \cdot (t - 2) - 4.9 \cdot (t - 2)^2$$

Notice that we denote y_f by y_2.

c)

In order to answer this question, it is necessary to use position-time equations of the objects.

$$y_1 = 20 \cdot t - 4.9 \cdot t^2$$

$$y_2 = 30 \cdot (t - 2) - 4.9 \cdot (t - 2)^2$$

The objects meet each other when $y_1 = y_2$.

Hence,

$$20 \cdot t - 4.9 \cdot t^2 = 30 \cdot (t - 2) - 4.9 \cdot (t - 2)^2$$

or in expanded form

$$20 \cdot t - 4.9 \cdot t^2 = -4.9 \cdot t^2 + 49.6 \cdot t - 79.6$$

or in alternate form

$$79.6 - 29.6 \cdot t = 0$$

We can solve for t

$$t \approx 2.69 \, s$$

Now, we can find where they will meet.

The position of the first object as a function of time is given by

$$y_1 = 20 \cdot t - 4.9 \cdot t^2$$

We substitute t into the equation and solve for y_1

$$y_1 = 20 \cdot 2.69 - 4.9 \cdot 2.69^2 \approx 18.34 \, m$$

d) The position-time graphs are parabolas.

$$y_1 = 20 \cdot t - 4.9 \cdot t^2$$

$$y_2 = 30 \cdot (t-2) - 4.9 \cdot (t-2)^2 = -4.9 \cdot t^2 + 49.6 \cdot t - 79.6$$

The both graphs have one common point, when the objects met.

We use the tables of values in order to draw the graphs. There are the physically meaningful values in the tables.

$y_1(m)$	$t(s)$	$y_2(m)$	$t(s)$
0	0		0
15.1	1		1
20.4	2	0	2
15.9	3	25.1	3
1.6	4	40.4	4
		45.9	5
		41.6	6

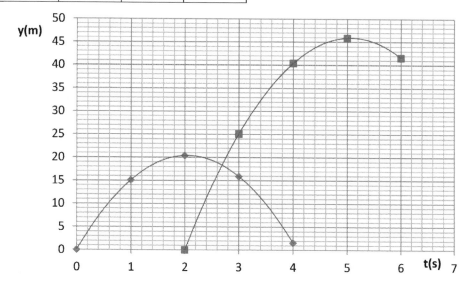

6.

a)

The equation for velocity as a function of time is given by

$$v_f = v_i + a \cdot t$$

where v_f - the final velocity of the object, v_i – the initial velocity of the object, a – the acceleration and t – the time it takes the object to change its velocity.

We rewrite the equation for the first stone by renaming v_i by v_1 and by setting $a = -g$.

$$v_f = v_1 - g \cdot t$$

First let's show that the first stone returns at the same magnitude of the velocity with which it left.

The velocity-position equation of motion of the object is given by

$$v_f{}^2 = v_i{}^2 + 2 \cdot a \cdot (y_f - y_i)$$

where y_f - the final position of the object, y_i – the initial position of the object, v_f - the final velocity of the object, v_i – the initial velocity of the object and a – the acceleration.

We rewrite the velocity-position equation of motion of the object by setting $a = -g$.

$$v_f{}^2 = v_i{}^2 - 2 \cdot g \cdot (y_f - y_i)$$

The initial position of the stone equals the final position of the stone $y_f = y_i$.

Hence,

$$v_f{}^2 = v_i{}^2 - 2 \cdot g \cdot 0$$

or

$$v_f{}^2 = v_i{}^2$$

From the last equation we can see that the final and initial velocities have equal magnitudes. However, we remember that the final and initial velocities have different directions.

Then, the final velocity of the first stone is

$$v_f = -v_i = -v_1$$

We can substitute it into the equation

$$v_f = v_1 - g \cdot t$$

and we get

$$-v_1 = v_1 - g \cdot t$$

or

$$-2v_1 = -g \cdot t \quad (1)$$

Now let's look on the second stone. It was dropped from a height. The equation for velocity as a function of time is given by

$$v_f = -g \cdot t$$

We rewrite the equation for the second stone by renaming v_f by v_2

$$v_2 = -g \cdot t \quad (2)$$

Finally, we can compare equation (1) with equation (2), we take into account that the both stones spend the same amount of time in the air and we get

$$v_2 = -2v_1$$

or

$$v_2 = -2 \cdot 20 = -40 \, \frac{m}{s}$$

b)

The velocity-position equation of motion of the second stone can be written as

$$v_f{}^2 = v_i{}^2 - 2 \cdot g \cdot (y_f - y_i)$$

21

where y_f - the final position of the stone, y_i – the initial position of the stone, v_f - the final velocity of the stone, v_i – the initial velocity of the stone and g – the acceleration due to gravity.

We substitute the values into the equation and solve for y_i

$$(-40)^2 = 0^2 - 2 \cdot 9.8 \cdot (0 - y_i)$$

$$y_i \approx 81.63 \ m$$

Finally, the height equals $81.63 \ m$.

c)

The velocity-time equation of the first stone is

$$v_{1f} = v_1 - g \cdot t$$

and the velocity-time equation of the second stone is

$$v_{2f} = -g \cdot t$$

We use the tables of values in order to draw the graphs.

$v_{1f}(\frac{m}{s})$	$t(s)$	$v_{2f}(\frac{m}{s})$	$t(s)$
20	0	0	0
10.2	1	-9.8	1
0.4	2	-19.6	2
-9.4	3	-29.4	3
-19.2	4	-39.2	4

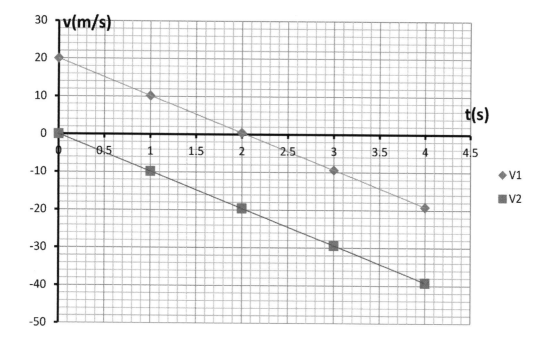

7.

a)

Since the sandbag has been moving with the hot air balloon, it already possesses the initial

vertical velocity of 5 $\frac{m}{s}$.

The equation for velocity as a function of time is given by

$$v_f = v_i - g \cdot t$$

Hence,

$$v_f = 5 - 9.8 \cdot 12 = -112.6 \, \frac{m}{s}$$

Finally, the speed of the sandbag when it hit the ground is $112.6 \, \frac{m}{s}$.

b)

The position-time equation of motion of the sandbag is given by

$$y_f = y_i + v_i \cdot t - \frac{g \cdot t^2}{2}$$

We know that the sandbag hit the ground, so $y_f = 0$. Also, it was in the air $12\ seconds$ and

it's initial vertical velocity is $5\ \frac{m}{s}$.

Thereby,

$$0 = y_i + 5 \cdot 12 - \frac{9.8 \cdot 12^2}{2}$$

$$y_i = 645.6\ m$$

Finally, the balloon was at a height of $645.6\ m$.

c)

The velocity-position equation of motion of the sandbag is given by

$$v_f{}^2 = v_i{}^2 - 2 \cdot g \cdot (y_f - y_i)$$

The sandbag will come to a complete stop for an instant when it reaches its maximum

height. It means that $v_f = 0$. Also, we use that $v_i = 5\ \frac{m}{s}$ and $y_i = 645.6\ m$.

$$0^2 = 5^2 - 2 \cdot 9.8 \cdot (y_f - 645.6)$$

$$y_f \approx 646.88\ m$$

Thereby, the sandbag will reach a height of $646.88\ m$.

d)

We calculate the velocity of the sandbag $6\ seconds$ after it was released.

The equation for velocity as a function of time is given by

$$v_f = v_i - g \cdot t$$

Hence,

$$v_f = 5 - 9.8 \cdot 6 = -53.8 \, \frac{m}{s}$$

The relative velocity of the sandbag with respect to the balloon is given by

$$v_{SB} = v_S - v_B$$

where v_{SB} – the velocity of the sandbag with respect to the balloon, v_S – the absolute velocity of the sandbag and v_B – the absolute velocity of the balloon.

Now, we substitute the values into the formula and solve

$$v_{SB} = -53.8 - 5.5 = -59.3 \, \frac{m}{s}$$

8.

a)

The initial velocity is the velocity at which motion begins.

Hence, the initial velocity is $30 \, \frac{m}{s}$ upwards.

b)

We can calculate acceleration by using the fact that the acceleration is equal to the slope of the velocity versus time graph.

$$a = \frac{\Delta v}{\Delta t} = \frac{v_f - v_i}{t_f - t_i}$$

We substitute the values into the formula and solve

$$a = \frac{v_f - v_i}{t_f - t_i} = \frac{-30 - 30}{10 - 0} = -6 \, \frac{m}{s^2}$$

Hence, the magnitude of the free-fall acceleration on the planet is $6 \, \frac{m}{s^2}$.

c)

The velocity-position equation of motion of the stone is given by

$$v_f^2 = v_i^2 - 2 \cdot a \cdot (y_f - y_i)$$

The stone will come to a complete stop for an instant when it reaches its maximum height.

It means that $v_f = 0$. Also, we use that $v_i = 30 \frac{m}{s}$, $y_i = 0\ m$ and $a = 6\frac{m}{s^2}$.

$$0^2 = 30^2 - 2 \cdot 6 \cdot (y_f - 0)$$

$$y_f = 75\ m$$

Also, we can answer the question by using the fact that displacement is the area between the function and the t-axis. So, we find the area of a triangle by multiplying the base by the height and dividing by two

$$\Delta y = \frac{30 \cdot 5}{2} = 75\ m$$

d)

The acceleration-time graph consists of a horizontal line, shows constant acceleration of the motion.

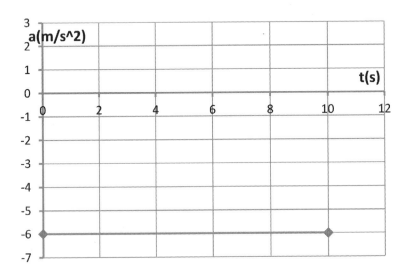

A Horizontally Launched Projectile.

An object is thrown (projected) horizontally to the earth's surface has horizontal (x-direction) and vertical (y-direction) motion. The horizontal and vertical motions can be analysed separately. The horizontal motion has a constant velocity v_{ix} and the vertical motion has a constant acceleration g. The horizontal velocity does not affect the vertical velocity. We can apply the kinematic equations to each axis independently.

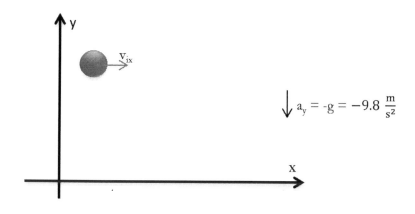

9.

a)

The position-time equation of vertical motion of the object is given by

$$y_f = y_i + v_{iy} \cdot t - \frac{g \cdot t^2}{2}$$

where y_f - the final vertical position of the object, y_i – the initial vertical position of the object, v_{iy} – the initial vertical velocity of the object, g – the acceleration due to gravity and t – the time it takes the object to move from the initial position to the final position.

The object is dropped from the airplane which travels horizontally, so its initial vertical velocity is zero ($v_{iy} = 0$).

Hence,

$$y_f = y_i - \frac{g \cdot t^2}{2}$$

We substitute the values into the equation and solve for y_i

$$0 = y_i - \frac{9.8 \cdot 18.6^2}{2}$$

$$y_i \approx 1695.2 \ m$$

b)

The equation for vertical velocity as a function of time is given by

$$v_{fy} = v_{iy} - g \cdot t$$

where v_{fy} - the final vertical velocity of the object, v_{iy} – the initial vertical velocity of the object, g – the acceleration due to gravity and t – the time it takes the object to change its vertical velocity.

We know that $-g = -9.8 \ \frac{m}{s^2}$, $v_{iy} = 0$ and $t = 18.6 \ s$.

Hence,

$$v_{fy} = 0 - 9.8 \cdot 18.6 = -182.28 \ \frac{m}{s}$$

c)

Since the object has been moving with the airplane, it already possesses the initial horizontal velocity of $125 \ \frac{m}{s}$. For horizontally launched projectile, the horizontal velocity of a projectile cannot change. Therefore,

$$v_{fx} = v_{ix} = 125 \ \frac{m}{s}$$

where v_{fx} - the final horizontal velocity of the object and v_{iy} – the initial horizontal velocity of the object.

28

d)

The following equations give the relationships between the magnitude and direction of velocity (v and θ) and its components (v_x and v_y)

$$v = \sqrt{v_x{}^2 + v_y{}^2}$$

$$\theta = tan^{-1}\left(\frac{v_y}{v_x}\right)$$

In our case, the equations will take the following forms

$$v_f = \sqrt{v_{fx}{}^2 + v_{fy}{}^2}$$

$$\theta_f = tan^{-1}\left(\frac{v_{fy}}{v_{fx}}\right)$$

where v_f - the magnitude of the final velocity and θ_f – the direction of the final velocity.

We substitute the values into the equations and solve

$$v_f = \sqrt{125^2 + (-182.28)^2} \approx 221.02\,\frac{m}{s}$$

$$\theta_f = tan^{-1}\left(\frac{-182.28}{125}\right) \approx -55.56^0\ with\ respect\ to\ the\ positive\ x - axis.$$

e)

First, we can use the position-time equation of vertical motion

$$y_f = y_i - \frac{g \cdot t^2}{2}$$

and calculate the vertical position of the object after the first 4 $seconds$ and the vertical position of the object after the first 5 $seconds$

$$y_4 = 1695.2 - \frac{9.8 \cdot 4^2}{2} = 1616.8\ m$$

$$y_5 = 1695.2 - \frac{9.8 \cdot 5^2}{2} = 1572.7\ m$$

We denote y_f by y_4 and y_5 for convenience.

Next, we subtract the vertical position of the object after the first $5\ seconds$ from the position of the object after the first $4\ seconds$

$$1572.7 - 1616.8 = -44.1\ \text{m}$$

Finally, we got that the object fell downwards **44.1 m** during the fifth second.

10.

a)

The position-time equation of horizontal motion of the tennis ball is given by

$$x_f = x_i + v_{ix} \cdot t$$

where x_f - the final horizontal position of the tennis ball, x_i – the initial horizontal position of the tennis ball, v_{ix} – the initial horizontal velocity of the tennis ball and t – the time it takes the tennis ball to move from the initial position to the final position.

We know that $x_i = 0$, $x_f = 12\ m$ and $v_{ix} = 24\ \frac{m}{s}$.

Hence,

$$12 = 0 + 24 \cdot t$$

$$t = 0.5\ s$$

It takes a half second for the tennis ball to cover the distance between the server and the net.

b)

We need to find the height of the tennis ball when it approached the net.

The position-time equation of vertical motion of the tennis ball is given by

$$y_f = y_i + v_{iy} \cdot t - \frac{g \cdot t^2}{2}$$

The tennis ball is served horizontally, so its initial vertical velocity is zero ($v_{iy} = 0$).

We know that $y_i = 2.7\ m$, $g = 9.8\ \frac{m}{s^2}$ and $t = 0.5\ s$.

30

We substitute the values into the equation and solve for y_f

$$y_f = 2.7 + 0 \cdot 0.5 - \frac{9.8 \cdot 0.5^2}{2}$$

$$y_f = 1.475 \ m$$

We can see that the tennis ball clears the net because the net is $1 \ m$ high. We can calculate the distance by finding the difference between the height of the tennis ball when it approached the net and the height of the net.

Finally,

$$1.475 - 1 = 0.475 \ m$$

11.

a)

The position-time equation of vertical motion of the stone is given by

$$y_f = y_i - \frac{g \cdot t^2}{2}$$

The stone was thrown from the top of the taller tower and cleared the smaller tower. By knowing this, we can calculate the time required for the stone to move between the two towers. We substitute the values into the equation and solve for t

$$40 = 50 - \frac{9.8 \cdot t^2}{2}$$

$$t \approx 1.43 \ s$$

Now we can find the distance between the two towers. The position-time equation of horizontal motion of the stone is given by

$$x_f = x_i + v_{ix} \cdot t$$

We know that $x_i = 0$ and $v_{ix} = 25\,\frac{m}{s}$. Hence,

$$x_f = 0 + 25 \cdot 1.43 = 35.75\,m$$

Since, the distance between the two towers is the horizontal displacement of the stone.

Finally,

$$\Delta x = x_f - x_i = 35.75 - 0 = 35.75\,m$$

b)

First, we calculate the total time the stone was in the air. We substitute the values into the

position-time equation of vertical motion and solve for t

$$y_f = y_i - \frac{g \cdot t^2}{2}$$

$$0 = 50 - \frac{9.8 \cdot t^2}{2}$$

$$t \approx 3.19\,s$$

Secondly, we can calculate the time required for the stone to reach the ground from the

smaller tower by subtracting the time required for the stone to move between the two

towers from the total time.

$$3.19 - 1.43 = 1.76\,s$$

Thirdly, we can calculate the position of the point on the ground where the stone lands by

using the position-time equation of horizontal motion

$$x_f = x_i + v_x \cdot t$$

We know that $x_i = 35.75\,m$, $t = 1.76\,s$ and $v_x = v_{ix} = 25\,\frac{m}{s}$.

Hence,

$$x_f = 35.75 + 25 \cdot 1.76 = 79.75\,m$$

Finally, we calculate the distance between the smaller tower and the point on the ground where the stone lands.

$$\Delta x = x_f - x_i = 79.75 - 35.75 = 44 \, m$$

c)

First, we write a formula for the vertical position of the stone as a function of time. We use the position-time equation of vertical motion

$$y_f = y_i - \frac{g \cdot t^2}{2}$$

The initial vertical position of the stone is $y_i = 50 \, m$ and $g = 9.8 \frac{m}{s^2}$.

Thereby,

$$y_f = 50 - \frac{9.8 \cdot t^2}{2}$$

or

$$y = 50 - 4.9 \cdot t^2 \quad (1)$$

Secondly, we write an expression for the horizontal position of the stone as a function of time. We use the position-time equation of horizontal motion

$$x_f = x_i + v_{ix} \cdot t$$

The initial horizontal position of the stone is $x_i = 0$ and the horizontal velocity is $v_{ix} = 25 \frac{m}{s}$.

Thereby,

$$x_f = 0 + 25 \cdot t$$

or

$$x = 25 \cdot t$$

Thirdly, we rewrite the last expression as

$$t = \frac{x}{25}$$

and substitute into the formula (1) for the vertical position of the stone as a function of time. We got the equation of the trajectory

$$y = 50 - 4.9 \cdot \left(\frac{x}{25}\right)^2$$

or

$$y = 50 - 0.00784 \cdot x^2$$

Fourthly, we calculate the points of the flight path using the equation of the trajectory. The values of the graph is rounded to the nearest integer.

$x(m)$	$y(m)$
0	50
10	49
20	47
30	43
40	37
50	30
60	22
70	12
80	0

Lastly, we got

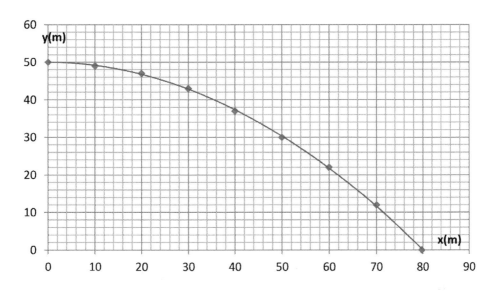

12.

First, we determine velocity v as a function of X and h.

The position-time equation of vertical motion of the stone is given by

$$y_f = y_i - \frac{g \cdot t^2}{2}$$

The first stone was thrown from the top of the taller tower. We know that $y_f = 0$ and $y_i = 4h$. Therefore,

$$0 = 4h - \frac{g \cdot t^2}{2}$$

or

$$t = \sqrt{\frac{8h}{g}}$$

The position-time equation of horizontal motion of the stone is given by

$$x_f = x_i + v_{ix} \cdot t$$

35

We know that for the first stone $x_i = 0$, $x_f = \frac{3}{4}X$, $v_{ix} = v$ and $t = \sqrt{\frac{8h}{g}}$.

Hence,

$$\frac{3}{4}X = v \cdot \sqrt{\frac{8h}{g}}$$

or

$$v = \frac{3}{4}X \cdot \sqrt{\frac{g}{8h}} \quad (1)$$

Secondly, we determine velocity u as a function of X and h.

The position-time equation of vertical motion of the stone is given by

$$y_f = y_i - \frac{g \cdot t^2}{2}$$

The second stone was thrown from the top of the smaller tower. We know that $y_f = 0$ and $y_i = h$. Therefore,

$$0 = h - \frac{g \cdot t^2}{2}$$

or

$$t = \sqrt{\frac{2h}{g}}$$

The position-time equation of horizontal motion of the stone is given by

$$x_f = x_i + v_{ix} \cdot t$$

We know that for the second stone $x_i = X$, $x_f = \frac{3}{4}X$, $v_{ix} = -u$ and $t = \sqrt{\frac{2h}{g}}$.

The initial horizontal velocity is negative, because the second stone moved in the direction opposite to the direction of the first stone.

Hence,

$$\frac{3}{4}X = X - u \cdot \sqrt{\frac{2h}{g}}$$

or

$$u = \frac{1}{4}X \cdot \sqrt{\frac{g}{2h}} \quad (2)$$

Thirdly, we divide the equation (2) by the equation (1).

$$\frac{u}{v} = \left(\frac{1}{4}X \cdot \sqrt{\frac{g}{2h}}\right) \div \left(\frac{3}{4}X \cdot \sqrt{\frac{g}{8h}}\right)$$

or

$$\frac{u}{v} = \left(\frac{1}{4}X \cdot \sqrt{\frac{g}{2h}}\right) \times \left(\frac{4}{3X} \cdot \sqrt{\frac{8h}{g}}\right)$$

Lastly, we simplify the expression and find

$$\frac{u}{v} = \frac{2}{3}$$

or

$$3u = 2v$$

13.

a)

The position-time equation of vertical motion of the bomb is given by

$$y_f = y_i - \frac{g \cdot t^2}{2}$$

We can calculate the time required for the bomb to hit the target. We substitute the values into the equation and solve for t

37

$$0 = 1000 - \frac{9.8 \cdot t^2}{2}$$

$$t \approx 14.29 \ s$$

The bomb is approaching the moving target, so we can use the relative velocity method. The method allows us to calculate the horizontal velocity of the bomb with respect to the target regarded as being at rest. The relative horizontal velocity of the bomb with respect to the target is given by

$$v_{BT} = v_B - v_T$$

where v_{BT} – the horizontal velocity of the bomb with respect to the target regarded as being at rest, v_B – the absolute horizontal velocity of the bomb and v_T – the absolute velocity of the target.

Hence,

$$v_{BT} = 50 - 20 = 30 \ \frac{m}{s}$$

This is how fast the bomb is approaching the target (horizontally).

In order to calculate the horizontal distance of the bomber from the target we use the position-time equation of horizontal motion of the bomb

$$x_f = x_i + v_{BT} \cdot t$$

Since, the horizontal distance is the horizontal displacement of the bomb we rewrite it as

$$\Delta x = x_f - x_i = v_{BT} \cdot t$$

Now we substitute the values into the equation and solve

$$\Delta x = v_{BT} \cdot t = 30 \cdot 14.29 = 428.7 \ m$$

b)

We draw the figure

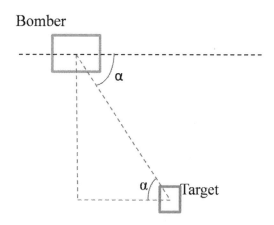

and calculate the angle using the tangent.

$$\tan(\alpha) = \frac{1000}{428.7}$$

$$\alpha = 66.8^0$$

c)

The equation for vertical velocity as a function of time is given by

$$v_{fy} = v_{iy} - g \cdot t$$

We know that $-g = -9.8 \, \frac{m}{s^2}$, $v_{iy} = 0$ and $t = 14.29 \, s$.

Hence,

$$v_{fy} = 0 - 9.8 \cdot 14.29 \approx -140.04 \, \frac{m}{s}$$

Since the bomb has been moving with the bomber, it already possesses the initial horizontal velocity of $50 \, \frac{m}{s}$. For horizontally launched projectile, the horizontal velocity of the projectile cannot change.

Therefore,

$$v_{fx} = v_{ix} = 50 \ \frac{m}{s}$$

The following equations give the relationships between the magnitude and direction of velocity (v_f and θ_f) and its components (v_{fx} and v_{fy})

$$v_f = \sqrt{v_{fx}^2 + v_{fy}^2}$$

$$\theta_f = tan^{-1}\left(\frac{v_{fy}}{v_{fx}}\right)$$

We substitute the values into the equations and solve

$$v_f = \sqrt{50^2 + (-140.04)^2} \approx 148.70 \ \frac{m}{s}$$

$$\theta_f = tan^{-1}\left(\frac{-140.04}{50}\right) \approx -70.35^0 \ with \ respect \ to \ the \ positive \ x - axis.$$

14.

a)

First, we write an equation for the horizontal position of the apple as a function of time. We use the position-time equation of horizontal motion

$$x_f = x_i + v_{ix} \cdot t$$

The initial horizontal position of the apple is $x_i = 0$ and the horizontal velocity is $v_{ix} = 10\frac{m}{s}$.

Thereby,

$$x_f = 0 + 10 \cdot t$$

or

$$x_{apple} = 10 \cdot t \quad (1)$$

Secondly, we write an equation for the vertical position of the apple as a function of time.

We use the position-time equation of vertical motion

$$y_f = y_i - \frac{g \cdot t^2}{2}$$

The initial vertical position of the apple is $y_i = 75 \ m$ and $g = 9.8 \frac{m}{s^2}$.

Thereby,

$$y_f = 75 - \frac{9.8 \cdot t^2}{2}$$

or

$$y_{apple} = 75 - 4.9 \cdot t^2 \quad (2)$$

Thirdly, the position-time equation of motion of the arrow is given by

$$y_f = y_i + v_i \cdot t - \frac{g \cdot t^2}{2}$$

We know that the initial position is $0 \ m$ and the initial velocity of the arrow is $30 \ \frac{m}{s}$.

Also, we can write that the time is $(t - 1)$ because the arrow was shot one second later.

Then,

$$y_{arrow} = 0 + 30 \cdot (t - 1) - \frac{9.8 \cdot (t - 1)^2}{2}$$

or

$$y_{arrow} = 30 \cdot (t - 1) - 4.9 \cdot (t - 1)^2 \quad (3)$$

Fourthly, we use the fact that the arrow hit the apple. It means that there exists a point

where $y_{arrow} = y_{apple}$ and $x_{arrow} = x_{apple}$.

We equate equations (2) and (3) and solve for t.

$$75 - 4.9 \cdot t^2 = 30 \cdot (t - 1) - 4.9 \cdot (t - 1)^2$$

$$75 - 4.9 \cdot t^2 = 30 \cdot t - 30 - 4.9 \cdot t^2 + 9.8 \cdot t - 4.9$$

$$75 - 4.9 \cdot t^2 = -4.9 \cdot t^2 + 39.8 \cdot t - 34.9$$

$$109.9 = 39.8 \cdot t$$

$$t \approx 2.76 \ s$$

Now we use the equation for the horizontal position of the apple.

$$x_{apple} = 10 \cdot t \quad (1)$$

We substitute $t = 2.76 \ s$ in (1) and obtain

$$x_{apple} = 10 \cdot 2.76 = 27.6 \ m$$

Lastly, the horizontal distance between the building and the archer equals to horizontal displacement of the apple.

$$\Delta x = x_{apple} - x_i = 27.6 - 0 = 27.6 \ m$$

b)

First, we find the velocity (magnitude and direction) of the apple immediately before the impact.

The equation for vertical velocity as a function of time is given by

$$v_{fy} = v_{iy} - g \cdot t$$

We know that $v_{iy} = 0$ and $t = 2.76 \ s$.

Hence,

$$v_{fy} = 0 - 9.8 \cdot 2.76 \approx -27.05 \ \frac{m}{s}$$

The horizontal velocity is constant.

$$v_{fx} = v_{ix} = 10 \ \frac{m}{s}$$

The following equations give the relationships between the magnitude and direction of velocity (v_f and θ_f) and its components (v_{fx} and v_{fy})

$$v_f = \sqrt{v_{fx}^2 + v_{fy}^2}$$

$$\theta_f = tan^{-1}\left(\frac{v_{fy}}{v_{fx}}\right)$$

We substitute the values into the equations and calculate the velocity (magnitude and direction) of the apple.

$$v_f = \sqrt{10^2 + (-27.05)^2} \approx 28.84\,\frac{m}{s}$$

$$\theta_f = tan^{-1}\left(\frac{-27.05}{10}\right) \approx -69.71^0 \text{ with respect to the positive } x - axis.$$

Now we find the velocity (magnitude and direction) of the arrow immediately before the impact.

The equation for vertical velocity as a function of time is given by

$$v_{fy} = v_{iy} - g \cdot t$$

We know that $v_{iy} = 30\,\frac{m}{s}$ and $t = 2.76\ s$.

Hence,

$$v_{fy} = 30 - 9.8 \cdot 2.76 \approx 2.95\,\frac{m}{s}$$

The direction of the velocity is positive (upwards).

A Projectile Launched At An Angle.

An object is projected at an angle to the horizontal in a gravitational field follows a parabolic path. It is necessary to consider the motion in a horizontal and in a vertical direction separately. The horizontal motion has a constant velocity v_{ix} and the vertical motion has a constant acceleration g. We can apply the kinematic equations to each axis independently.

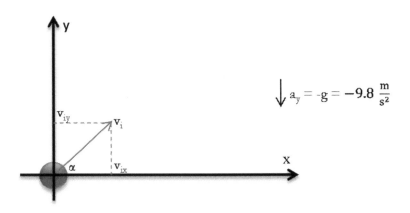

15.

a)

The position-time equation of horizontal motion of the soccer ball is given by

$$x_f = x_i + v_{ix} \cdot t$$

where x_f - the final horizontal position of the soccer ball, x_i – the initial horizontal position of the soccer ball, v_{ix} – the initial horizontal velocity of the soccer ball and t – the time it takes the soccer ball to move from the initial position to the final position.

The formula for the initial horizontal velocity is

$$v_{ix} = v_i \cdot \cos(\alpha)$$

where v_i – the initial velocity of the soccer ball and α - the angle that the velocity vector makes with the horizontal.

Thereby, the position-time equation can be written as

$$x_f = x_i + v_i \cdot \cos(\alpha) \cdot t$$

We assume that the initial horizontal position is $0\ m$ and we know that the initial velocity is

$20\ \frac{m}{s}$ and α is 40^0.

Then,

$$x_f = 0 + 20 \cdot \cos(40) \cdot t$$

or

$$x = 15.32 \cdot t$$

Notice that we denote x_f by x.

b)

The position-time equation of vertical motion of the soccer ball is given by

$$y_f = y_i + v_{iy} \cdot t - \frac{g \cdot t^2}{2}$$

where y_f - the final vertical position of the soccer ball, y_i – the initial vertical position of the

soccer ball, v_{iy} – the initial vertical velocity of the soccer ball, g – the acceleration due to

gravity and t – the time it takes the soccer ball to move from the initial position to the final

position.

The formula for the initial vertical velocity is

$$v_{iy} = v_i \cdot \sin(\alpha)$$

where v_i – the initial velocity of the soccer ball and α - the angle that the velocity vector

makes with the horizontal.

Thereby, the position-time equation can be written as

$$y_f = y_i + v_i \cdot \sin(\alpha) \cdot t - \frac{g \cdot t^2}{2}$$

The soccer ball is kicked from the ground, so its initial vertical position is $0\ m$. Also, the

initial velocity is $20\ \frac{m}{s}$, α is 40^0 and $g = 9.8\ \frac{m}{s^2}$.

Hence,

$$y_f = 0 + 20 \cdot \sin(40) \cdot t - \frac{9.8 \cdot t^2}{2}$$

or

$$y = 12.86 \cdot t - 4.9 \cdot t^2$$

Notice that we denote y_f by y.

c)

We rewrite position-time equation of horizontal motion as

$$t = \frac{x}{15.32}$$

and substitute it into the position-time equation of vertical motion

$$y = 12.86 \cdot t - 4.9 \cdot t^2$$

We got the equation of the trajectory

$$y = 12.86 \cdot \frac{x}{15.32} - 4.9 \cdot \left(\frac{x}{15.32}\right)^2$$

or

$$y = 0.84 \cdot x - 0.02 \cdot x^2$$

d)

We use the position-time equation of vertical motion of the soccer ball

$$y = 12.86 \cdot t - 4.9 \cdot t^2$$

We substitute $y = 0$ into the equation and solve for t

$$0 = 12.86 \cdot t - 4.9 \cdot t^2$$

There are two solutions for this equation. First, $t = 0$ gives the time when the soccer ball starts to move. Second, $t \approx 2.62\ s$ gives the time while the ball was in the air.

e)

We use the position-time equation of horizontal motion

$$x = 15.32 \cdot t$$

We substitute $t = 2.62\ s$ into the equation and solve for x

$$x = 15.32 \cdot 2.62 = 40.14\ m$$

16.

a)

First, we write the position-time equation of horizontal motion of the projectile.

$$x_f = x_i + v_i \cdot \cos(\alpha) \cdot t$$

We know that $x_f = 700\ m$, $x_i = 0$ and $t = 10\ s$.

Hence,

$$700 = v_i \cdot \cos(\alpha) \cdot 10$$

or

$$70 = v_i \cdot \cos(\alpha) \quad (1)$$

Secondly, we write the position-time equation of vertical motion

$$y_f = y_i + v_i \cdot \sin(\alpha) \cdot t - \frac{g \cdot t^2}{2}$$

We use that $y_f = 90\ m$, $y_i = 0$, $t = 10\ s$ and $g = 9.8\frac{m}{s^2}$.

Hence,

$$90 = v_i \cdot \sin(\alpha) \cdot 10 - \frac{9.8 \cdot 10^2}{2}$$

or

$$58 = v_i \cdot \sin(\alpha) \quad (2)$$

Thirdly, we divide the (2) equation by the (1) equation. We obtain

$$\tan(\alpha) = \frac{58}{70}$$

or

$$\alpha = 39.64^0 \ with \ respect \ to \ the \ positive \ x - axis.$$

Lastly, we substitute $\alpha = 39.64^0$ into the (2) equation and solve for v_i

$$58 = v_i \cdot \sin(39.64)$$

$$v_i \approx 90.91 \ \frac{m}{s}$$

b)

The equation for vertical velocity as a function of time is given by

$$v_{fy} = v_{iy} - g \cdot t$$

and the formula for the initial vertical velocity is

$$v_{iy} = v_i \cdot \sin(\alpha)$$

Thereby,

$$v_{fy} = v_i \cdot \sin(\alpha) - g \cdot t$$

The projectile reaches its maximum height when the vertical component of the velocity is zero. Hence,

$$0 = v_i \cdot \sin(\alpha) - g \cdot t$$

We substitute the values into the equation and solve for t

$$0 = 90.91 \cdot \sin(39.64) - 9.8 \cdot t$$

$$t \approx 5.92 \ s$$

Now, we use the position-time equation of vertical motion to calculate the maximum height reached by the projectile.

$$y_f = y_i + v_i \cdot \sin(\alpha) \cdot t - \frac{g \cdot t^2}{2}$$

We know that $y_i = 0$, $v_i = 90.91 \frac{m}{s}$, $\alpha = 39.64^0$, $t = 5.92\ s$ and $g = 9.8 \frac{m}{s^2}$.

Hence,

$$y_f = 0 + 90.91 \cdot \sin(39.64) \cdot 5.92 - \frac{9.8 \cdot 5.92^2}{2}$$

$$y_f \approx 171.62\ m$$

c)

The equation for vertical velocity as a function of time is given by

$$v_{fy} = v_i \cdot \sin(\alpha) - g \cdot t$$

We know that $v_i = 90.91 \frac{m}{s}$, $\alpha = 39.64^0$ and $t = 10\ s$.

Hence,

$$v_{fy} = 90.91 \cdot \sin(39.64) - 9.8 \cdot 10 = -40 \frac{m}{s}$$

The horizontal velocity is constant and formula is

$$v_{fx} = v_{ix} = v_i \cdot \cos(\alpha)$$

We substitute the values and calculate

$$v_{fx} = v_{ix} = 90.91 \cdot \cos(39.64) = 70 \frac{m}{s}$$

The following equations give the relationships between the magnitude and direction of velocity (v_f and θ_f) and its components (v_{fx} and v_{fy})

$$v_f = \sqrt{v_{fx}^2 + v_{fy}^2}$$

$$\theta_f = tan^{-1}\left(\frac{v_{fy}}{v_{fx}}\right)$$

We substitute the values into the equations and calculate the velocity (magnitude and direction) of the projectile.

$$v_f = \sqrt{70^2 + (-40)^2} \approx 80.62 \, \frac{m}{s}$$

$$\theta_f = tan^{-1}\left(\frac{-4}{70}\right) \approx -29.74^0 \; with \; respect \; to \; the \; positive \; x - axis.$$

17.

a)

The position-time equation of vertical motion of the projectile is given by

$$y_f = y_i + v_i \cdot \sin(\alpha) \cdot t - \frac{g \cdot t^2}{2}$$

and we know that $y_f - y_i = 0$. (The projectile starts and ends at the same y-position.)

Hence,

$$0 = v_i \cdot \sin(\alpha) \cdot t - \frac{g \cdot t^2}{2}$$

We solve the equation for t. There are two solutions for this equation. First, $t = 0$ gives the time when the projectile starts to move. Second,

$$t = \frac{2 \cdot v_i \cdot \sin(\alpha)}{g} \quad (1)$$

gives the time while the projectile was in the air.

The position-time equation of horizontal motion of the projectile is given by

$$x_f = x_i + v_i \cdot \cos(\alpha) \cdot t$$

and displacement, in our case the horizontal range R, is the shortest distance between initial and final horizontal positions and is given by the formula

$$R = x_f - x_i$$

Hence,

$$R = v_i \cdot \cos(\alpha) \cdot t \quad (2)$$

We substitute expression (1) into (2) and we get

$$R = \frac{2 \cdot \sin(\alpha) \cdot \cos(\alpha) \cdot v_i^2}{g}$$

Now we can use the double-angle identity for $\sin$

$$\sin(2\alpha) = 2 \cdot \sin(\alpha) \cdot \cos(\alpha)$$

and finally we get

$$R = \frac{v_i^2 \cdot \sin(2\alpha)}{g}$$

b)

The horizontal distance (range) would be greatest when $\sin(2\alpha)$ is greatest.

Hence,

$$\sin(2\alpha) = 1$$

$$2\alpha = 90^0$$

$$\alpha = 45^0$$

c)

The vertical velocity-position equation of motion of the projectile can be written as

$$v_{fy}^2 = v_{iy}^2 - 2 \cdot g \cdot (y_f - y_i)$$

51

where y_f - the final vertical position of the projectile, y_i – the initial vertical position of the projectile, v_{fy} - the final vertical velocity of the projectile, v_{iy} – the initial vertical velocity of the projectile and g – the acceleration due to gravity.

The formula for the initial vertical velocity is

$$v_{iy} = v_i \cdot \sin(\alpha)$$

and the projectile reaches its maximum height when the vertical component of the velocity is zero $v_{fy} = 0$.

Hence,

$$0 = (v_i \cdot \sin(\alpha))^2 - 2 \cdot g \cdot (y_f - y_i) \quad (1)$$

The displacement, in our case the maximum height H, is the shortest distance between initial and final vertical positions and is given by the formula

$$H = y_f - y_i \quad (2)$$

We rewrite equation (1) using (2)

$$0 = (v_i \cdot \sin(\alpha))^2 - 2 \cdot g \cdot H$$

or

$$H = \frac{v_i{}^2 \cdot \sin^2(\alpha)}{2 \cdot g}$$

18.

a)

First we will prove that the horizontal range of a projectile is given by $R = \frac{v_i{}^2 \cdot \sin(2\alpha)}{g}$.

The position-time equation of vertical motion of the projectile is given by

$$y_f = y_i + v_i \cdot \sin(\alpha) \cdot t - \frac{g \cdot t^2}{2}$$

and we know that $y_f - y_i = 0$. (The projectile starts and ends at the same y-position.)

Hence,

$$0 = v_i \cdot \sin(\alpha) \cdot t - \frac{g \cdot t^2}{2}$$

We solve the equation for t. There are two solutions for this equation. First, $t = 0$ gives the time when the projectile starts to move. Second,

$$t = \frac{2 \cdot v_i \cdot \sin(\alpha)}{g} \quad (1)$$

gives the time while the projectile was in the air.

The position-time equation of horizontal motion of the projectile is given by

$$x_f = x_i + v_i \cdot \cos(\alpha) \cdot t$$

and displacement, in our case the horizontal range R, is the shortest distance between initial and final horizontal positions and is given by the formula

$$R = x_f - x_i$$

Hence,

$$R = v_i \cdot \cos(\alpha) \cdot t \quad (2)$$

We substitute expression (1) into (2) and we get

$$R = \frac{2 \cdot \sin(\alpha) \cdot \cos(\alpha) \cdot v_i^2}{g}$$

Now we can use the double-angle identity for $\sin$

$$\sin(2\alpha) = 2 \cdot \sin(\alpha) \cdot \cos(\alpha)$$

and finally we get

$$R = \frac{v_i^2 \cdot \sin(2\alpha)}{g}$$

We know that for a given value of initial velocity, the horizontal distance (range) would be maximum at a launch angle of 45^0. We substitute the launch angle of 45^0 into the last formula and simplify.

$$R_{max} = \frac{v_i{}^2}{g}$$

Then, in order to calculate the initial velocity of the stone we substitute the maximum horizontal distance of $100\ m$ into the formula and we get

$$100 = \frac{v_i{}^2}{g}$$

or

$$100 = \frac{v_i{}^2}{9.8}$$

Finally,

$$v_i \approx 31.30\ \frac{m}{s}$$

We have dropped the negative value because it is not physically meaningful.

b)

First we will prove that maximum height of a projectile is given by $H = \frac{v_i{}^2 \cdot \sin^2(\alpha)}{2 \cdot g}$.

The vertical velocity-position equation of motion of the projectile can be written as

$$v_{fy}{}^2 = v_{iy}{}^2 - 2 \cdot g \cdot (y_f - y_i)$$

where y_f - the final vertical position of the projectile, y_i – the initial vertical position of the projectile, v_{fy} - the final vertical velocity of the projectile, v_{iy} – the initial vertical velocity of the projectile and g – the acceleration due to gravity.

The formula for the initial vertical velocity is

$$v_{iy} = v_i \cdot \sin(\alpha)$$

54

and the projectile reaches its maximum height when the vertical component of the velocity is zero $v_{fy} = 0$.

Hence,

$$0 = (v_i \cdot \sin(\alpha))^2 - 2 \cdot g \cdot (y_f - y_i) \quad (1)$$

The displacement, in our case the maximum height H, is the shortest distance between initial and final vertical positions and is given by the formula

$$H = y_f - y_i \quad (2)$$

We rewrite equation (1) using (2)

$$0 = (v_i \cdot \sin(\alpha))^2 - 2 \cdot g \cdot H$$

or

$$H = \frac{v_i^2 \cdot \sin^2(\alpha)}{2 \cdot g}$$

We know that for a given value of initial velocity, the horizontal distance (range) would be maximum at a launch angle of 45^0. Hence,

$$H = \frac{v_i^2}{4 \cdot g}$$

Also, we already proved that the maximum horizontal range of a projectile is given by

$$R_{max} = \frac{v_i^2}{g}$$

We compare the two last formulas and we get

$$R_{max} = 4H$$

Finally,

$$H = \frac{R_{max}}{4} = \frac{100}{4} = 25 \; m$$

Without even noticing, we proved that if the projectile is launched at 45^0 to the horizontal, then the maximum horizontal distance is four times the maximum height.

c)

The velocity-position equation of the stone, which was thrown up, is given by

$$v_f^2 = v_i^2 - 2 \cdot g \cdot (y_f - y_i)$$

The stone will come to a complete stop for an instant when it reaches its maximum height. It means that $v_f = 0$. Also, we use that

$$R_{max} = \frac{v_i^2}{g}$$

and $y_i = 0 \, m$.

Hence,

$$0 = R_{max} \cdot g - 2 \cdot g \cdot (y_f - 0)$$

or

$$0 = R_{max} - 2 \cdot y_f$$

and finally,

$$y_f = \frac{R_{max}}{2} = \frac{100}{2} = 50 \, m$$

It is important to mention that we proved

$$y_{f \, max} = \frac{R_{max}}{2}$$

It means that if we have the constant magnitude of the initial velocity then the maximum height of the vertically launched projectile equals to the half of the maximum horizontal distance of the projectile launched at an angle.

56

19.

a)

The position-time equation of vertical motion of the package is given by

$$y_f = y_i + v_i \cdot \sin(\alpha) \cdot t - \frac{g \cdot t^2}{2}$$

We know that $y_i = 1200\ m$, $y_f = 0$, $v_i = 80\ \frac{m}{s}$, $\alpha = -30^0$(below the horizontal) and $g = 9.8\frac{m}{s^2}$.

Hence,

$$0 = 1200 + 80 \cdot \sin(-30) \cdot t - \frac{9.8 \cdot t^2}{2}$$

$$t \approx 12.09\ s$$

We have dropped the negative value because it is not physically meaningful.

b)

The position-time equation of horizontal motion of the package is given by

$$x_f = x_i + v_i \cdot \cos(\alpha) \cdot t$$

We substitute the values into the equation and solve for x_f

$$x_f = 0 + 80 \cdot \cos(-30) \cdot 12.09$$

$$x_f \approx 837.62\ m$$

c)

The equation for vertical velocity as a function of time is given by

$$v_{fy} = v_i \cdot \sin(\alpha) - g \cdot t$$

We know that $v_i = 80\ \frac{m}{s}$, $\alpha = -30^0$ and $t = 12.09\ s$.

Hence,

$$v_{fy} = 80 \cdot \sin(-30) - 9.8 \cdot 12.09 \approx -158.48 \ \frac{m}{s}$$

The horizontal velocity is constant and formula is

$$v_{fx} = v_{ix} = v_i \cdot \cos(\alpha)$$

We substitute the values and calculate

$$v_{fx} = v_{ix} = 80 \cdot \cos(-30) \approx 69.28 \ \frac{m}{s}$$

The following equations give the relationships between the magnitude and direction of velocity (v_f and θ_f) and its components (v_{fx} and v_{fy})

$$v_f = \sqrt{v_{fx}^2 + v_{fy}^2}$$

$$\theta_f = tan^{-1} \left(\frac{v_{fy}}{v_{fx}}\right)$$

We substitute the values into the equations and calculate the velocity (magnitude and direction) of the package.

$$v_f = \sqrt{69.28^2 + (-158.48)^2} \approx 172.96 \ \frac{m}{s}$$

$$\theta_f = tan^{-1} \left(\frac{-158.48}{69.28}\right) \approx -66.39^0 \ with \ respect \ to \ the \ positive \ x - axis.$$

d)

First, we draw horizontal position-time graph.

The position-time equation of horizontal motion of the package is given by

$$x_f = x_i + v_i \cdot \cos(\alpha) \cdot t$$

Also, we know that $v_i \cdot \cos(\alpha) \approx 69.28 \ \frac{m}{s}$ and $x_i = 0$.

Then, the horizontal position of the package as a function of time is given by

$$x = 69.28 \cdot t$$

We use the table of values in order to draw the graph.

$x(m)$	$t(s)$
0	0
837.62	12.09

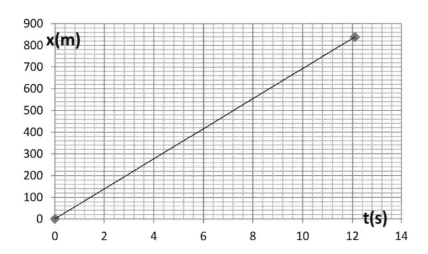

Now, we draw vertical position-time graph.

The position-time equation of vertical motion of the package is given by

$$y_f = y_i + v_i \cdot \sin(\alpha) \cdot t - \frac{g \cdot t^2}{2}$$

We know that $y_i = 1200\ m$, $v_i \cdot \sin(\alpha) = 80 \cdot \sin(-30) = -40\frac{m}{s}$ and $g = 9.8\ \frac{m}{s^2}$.

So, the vertical position of the package as a function of time is given by

$$y = 1200 - 40 \cdot t - 4.9 \cdot t^2$$

We use the table of values in order to draw the graph.

y(m)	t(s)
1200	0
961.6	4
566.4	8
310	10
0	12.09

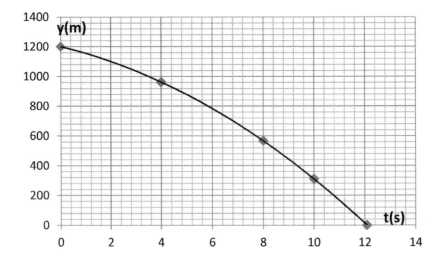

20.

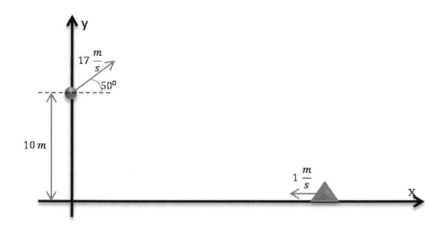

a)

The position-time equation of vertical motion of the package is given by

$$y_f = y_i + v_i \cdot \sin(\alpha) \cdot t - \frac{g \cdot t^2}{2}$$

We know that $y_i = 10\ m$, $y_f = 0$, $v_i = 17\ \frac{m}{s}$, $\alpha = 50^0$ and $g = 9.8 \frac{m}{s^2}$.

Hence,

$$0 = 10 + 17 \cdot \sin(50) \cdot t - \frac{9.8 \cdot t^2}{2}$$

$$t \approx 3.28\ s$$

We have dropped the negative value because it is not physically meaningful.

b)

The equation for vertical velocity as a function of time is given by

$$v_{fy} = v_i \cdot \sin(\alpha) - g \cdot t$$

We know that $v_i = 17\ \frac{m}{s}$, $\alpha = 50^0$ and $t = 3.28\ s$.

Hence,

$$v_{fy} = 17 \cdot \sin(50) - 9.8 \cdot 3.28 \approx -19.12 \; \frac{m}{s}$$

The horizontal velocity is constant and formula is

$$v_{fx} = v_{ix} = v_i \cdot \cos(\alpha)$$

We substitute the values and calculate

$$v_{fx} = v_{ix} = 17 \cdot \cos(50) \approx 10.93 \; \frac{m}{s}$$

The following equations give the relationships between the magnitude and direction of velocity ($v_f \; and \; \theta_f$) and its components ($v_{fx} \; and \; v_{fy}$)

$$v_f = \sqrt{v_{fx}^2 + v_{fy}^2}$$

$$\theta_f = tan^{-1}\left(\frac{v_{fy}}{v_{fx}}\right)$$

We substitute the values into the equations and calculate the velocity (magnitude and direction) of the package.

$$v_f = \sqrt{10.93^2 + (-19.12)^2} \approx 22.02 \, \frac{m}{s}$$

$$\theta_f = tan^{-1}\left(\frac{-19.12}{10.93}\right) \approx -60.25^0 \; with \; respect \; to \; the \; positive \; x - axis.$$

c)

The position-time equation of horizontal motion of the package is given by

$$x_f = x_i + v_i \cdot \cos(\alpha) \cdot t$$

Also, we know that $v_i \cdot \cos(\alpha) \approx 10.93 \; \frac{m}{s}$ and $x_i = 0$.

Then, the horizontal position of the package as a function of time is

$$x_{package} = 10.93 \cdot t$$

The position-time equation of motion of the ship is given by

$$x_f = x_i + v \cdot t$$

and we know that the ship has a negative velocity because it moves in the negative direction.

Thereby,

$$x_{ship} = x_i - 1 \cdot t$$

The package should land at the front of the ship. It means that $x_{package} = x_{ship}$ or

$$10.93 \cdot t = x_i - 1 \cdot t$$

$$11.93 \cdot t = x_i$$

The package was $3.28 \ seconds$ in the air. Hence,

$$x_i = 11.93 \cdot 3.28 \approx 39.13 \ m$$

The ship should be $39.13 \ m$ from the dock when the package is thrown.

d)

First, we draw horizontal velocity-time graph.

The horizontal motion has a constant velocity $v_x = v_{ix} = v_i \cdot \cos(\alpha) \approx 10.93 \ \frac{m}{s}$.

We use the table of values in order to draw the graph.

$v_x (\frac{m}{s})$	$t(s)$
10.93	0
10.93	3.28

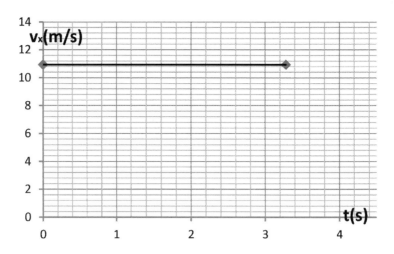

Now, we draw vertical velocity-time graph.

The equation for vertical velocity as a function of time is given by

$$v_{fy} = v_i \cdot \sin(\alpha) - g \cdot t$$

We know that $v_i = 17 \frac{m}{s}$, $\alpha = 50^0$ and $g = 9.8 \frac{m}{s^2}$.

So, the vertical velocity as a function of time is given by

$$v_y = 17 \cdot \sin(50) - 9.8 \cdot t$$

or

$$v_y = 13.02 - 9.8 \cdot t$$

We use the table of values in order to draw the graph.

$v_y(\frac{m}{s})$	$t(s)$
13.02	0
-19.12	3.28

64

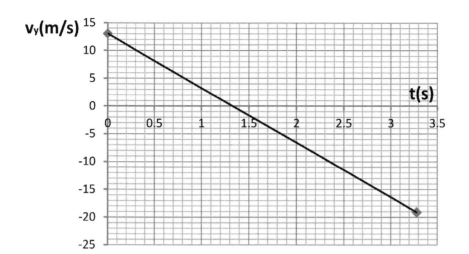

Dear Reader,

Thank you for learning with BABY STEPS IN PHYSICS Free-Fall and Projectile Motion. I hope you enjoy this book as much as my students do. This book is your third step in Physics and I believe that now you are ready for the next step.

Good Luck!

Best regards,

Boris Sapozhnikov

Printed in Great Britain
by Amazon

48226891R00041